DATE DUE			

599.88 Horton, Casey. 15356
HOR
           Apes

MESA VERDE MIDDLE SCHOOL
POWAY UNIFIED SCHOOL DISTRICT

443347 01823 50924C 51629F   006

ENDANGERED!

APES

Casey Horton

Series Consultant: James G. Doherty
General Curator, The Bronx Zoo, New York

BENCHMARK BOOKS

MARSHALL CAVENDISH

NEW YORK

Benchmark Books
Marshall Cavendish Corporation
99 White Plains Road
Tarrytown, New York 10591-9001

©Marshall Cavendish Corporation, 1996

All rights reserved. No part of this book may be reproduced or utilized in any form or by any means electronic or mechanical including photocopying, recording, or by any information storage and retrieval system, without permission from the copyright holders.

Library of Congress Cataloging-in-Publication Data

Horton, Casey.
    Apes / Casey Horton.
       p.   cm. — (Endangered!)
    Includes bibliographical references (p.   ) and index.
    ISBN 0-7614-0212-8
    1. Apes—Juvenile literature.   2. Endangered species—Juvenile literature.   [1. Apes. 2. Endangered species.]   I. Title.
    II. Series: Horton, Casey. Endangered!
QL737.P96H67 1995
599.88—dc20   95-12427
                                                                     CIP
                                                                       AC

Printed in Hong Kong

PICTURE CREDITS

The publishers would like to thank the following picture libraries for supplying the photographs used in this book: Ardea 16, 19; Bruce Coleman 26; Frank Lane Picture Agency FC, 2, 5, 6, 7, 10, 11, 12, 13, 14, 15, 17, 18, 20, 21, 22, 24, BC; Frank Lane Picture Agency/AGE Fotostock 8, 25; Frank Lane Picture Agency/Silvestris 9; Frank Lane Picture Agency/SUNSET 4, 23; Natural History Photographic Agency 27; Oxford Scientific Films 28, 29.

Series created by Brown Packaging

Front cover: Western lowland gorilla.
Title page: Common chimpanzees.
Back cover: Pygmy chimpanzee.

Contents

Introduction	4
Gorillas	5
Western Lowland Gorilla	8
Mountain Gorilla	10
Eastern Lowland Gorilla	12
Common Chimpanzee	14
Pygmy Chimpanzee	18
Orangutan	20
Gibbons	24
Kloss's & Javan Gibbons	26
Black Gibbon	28
Useful Addresses	30
Further Reading	30
Glossary	31
Index	32

Introduction

The apes are among the most intelligent animals on Earth. They are also the animals most closely related to human beings. In this book, we will look at many kinds of apes and find out how they live. Sadly, we will also discover that many of them are in danger of becoming **extinct**.

Since life first appeared on our planet millions of years ago, many kinds of animals have died out. They could not **adapt** to natural changes in their environment. Today's apes, however, are not threatened by natural changes, but by changes made by people. People are cutting down the apes' forest homes. The apes are running out of space.

Apes are divided into two groups: the lesser apes and the great apes. The lesser apes are the gibbons. The great apes are the chimpanzees, the orangutan, and the gorillas.

An orangutan stoops to take a drink of water. The orangutan is the only one of the great apes that lives in Asia. The gorillas and the chimps all live in Africa.

For many years people thought the gorilla was dangerous, because of its great size and its frightening appearance. Scientists now know that this is not true. The gorilla is normally a peaceful animal, and is often called "the gentle giant."

Gorillas

The gorillas are the largest of the apes. They are also the largest of the **primates**, the group of animals to which the apes belong. There are three different kinds of gorillas – the western lowland gorilla, the eastern lowland gorilla, and the mountain gorilla. They have much in common.

Most male gorillas weigh 300-400 lb (136-182 kg) and are up to 6 ft (1.8 m) tall standing on their hind legs. But some weigh as much as 485 lb (220 kg). The females are smaller. They weigh 150-300 lb (68-136 kg) and grow to a

Gorillas

height of 5 ft (1.5 m). Both males and females have a brownish gray coat and black skin. Adult males develop a silver-gray patch on their backs. This is why they are often known as **silverbacks**.

Gorillas live in parts of Central and West Africa, in open forests, where the trees are widely spaced. They feed on the leaves and stems of nettles, giant celery, and other plants. Some gorillas also eat termites when they find them, but gorillas are mainly **vegetarians**.

Gorillas spend less time in trees than the other apes do. In fact, the big males don't usually climb trees at all because they are too heavy. Females, however, like this western lowland gorilla, do sometimes climb trees, and young gorillas like to swing in the branches.

Gorillas live in groups called **troops**, which usually contain 10-15 animals. The head of the group is a very large, strong silverback male. There are normally several females in the troop, as well as their babies, and some young males.

Gorillas spend each morning searching for food and eating. At midday, they rest. Like most **mammals**, gorillas are covered in hair. This must be kept clean, so during the afternoon, they often **groom** each other. They look through each other's coats, removing insects, dirt, and knots of fur. At night, gorillas sleep in nests in the trees or on the ground.

The silverback protects the troop from danger. If he gets frightened or angry, he may bark, hoot, and beat his chest to scare off an enemy. He will rarely charge.

Areas where the western lowland gorilla can be found

The western lowland gorilla has an overhanging tip to its nose. Other gorillas have more rounded and open nostrils.

Western Lowland Gorilla

This gorilla lives in low-lying forests in parts of West and Central Africa. It is the smallest of the three kinds of gorillas. The adult male's silver-gray patch runs all the way over and down its back to its thighs. In the other two kinds of gorillas, the adult male's silver patch is on its back only.

The western lowland gorilla is the least rare of all the gorillas, but it is still at risk. Today there are about 40,000 in the wild. There used to be more. For example, at one time there were large numbers of this kind of gorilla in Nigeria. Now it is probably extinct there. The main reason for its disappearance is that people have destroyed its **habitat**. Most of the forests in Nigeria where the gorilla once made its home have been cut down. The land is now used for farms and ranches, and the gorilla has nowhere to live. In other parts of its **range**, the western lowland gorilla has been killed because it eats crops, such as bananas. Local people also hunt the gorilla and use its meat for food.

The countries in which the western lowland gorilla lives have laws against killing it. They have created national parks and **reserves** to protect the animals and their forest homes. But running these parks and reserves is very expensive. Often the governments don't have enough money to protect the gorillas completely.

A very young western lowland gorilla with its mother. A big problem for gorillas is that they breed very slowly. In some places, this can mean that not enough gorillas are being born to replace those being killed.

Area where the mountain gorilla can be found

Mountain gorillas in their misty forest home.

Mountain Gorilla

The mountain gorilla looks like the other gorillas, but its hair is much longer, especially on its arms and head. This is because the gorilla makes its home in the highlands, where it is very cold. It lives in just one small part of Central Africa: in Uganda's Bwindi Forest and in the Virunga Mountains, on the borders of Uganda, Rwanda, and Zaire.

The mountain gorilla is in great danger of dying out. Much of its forest habitat has been cleared for people and their farms. Also, many gorillas have been hunted and killed so that their hands could be sold as souvenirs to visitors. Scientists believe that only about 400 mountain gorillas are left in the Virunga Mountains. Between 80 and 130 remain in the Bwindi Forest.

Steps have been taken to look after these harmless giants. In Uganda, the Kigezi Gorilla Sanctuary has been created. This is a protected area in which mountain gorillas can live fairly peacefully and safely.

Meanwhile, in Rwanda, there is the Karisoke Research Centre, where scientists study mountain gorillas and their ways. It is located in the Virunga National Park. During a recent war in Rwanda, the guards at the park were forced to flee. There were fears that **poaching** (illegal hunting) would take place while they were away. Happily, the guards have now returned and found the gorillas unharmed.

A magnificent mountain silverback feeds on forest plants.

Area where the eastern lowland gorilla can be found

The eastern lowland gorilla survives in three national parks in eastern Zaire, in Central Africa.

Eastern Lowland Gorilla

The eastern lowland gorilla is the biggest of the gorillas and has the darkest coat. It lives in the forests of eastern Zaire. It is not as rare as the mountain gorilla, but it is still in serious danger of extinction. Like the other gorillas, the eastern lowland gorilla is losing its home as the forest is cut down to provide people with farmland.

Another problem facing the eastern lowland gorilla is that people still hunt it, even though it is against the law. As in the case of the mountain gorilla, hunters try to sell eastern lowland gorilla hands (and also skulls) to tourists as souvenirs. Also, some people in Zaire believe that a gorilla finger will act as a lucky charm. On top of that, some gorillas are injured by mistake when they become caught in traps meant for other animals.

However, **conservationists** are working to keep these gorillas safe. The Eastern Zaire Gorilla Conservation Project has been set up to study the animals' needs. But such projects are expensive. One plan is to ask tourists to pay to see the gorillas. This would raise some of the money needed to save these "gentle giants."

A family of eastern lowland gorillas feeds on bananas. Only about 10,000 of these apes survive.

Areas where the common chimpanzee can be found

A young chimpanzee practices climbing. Chimps are at home in trees.

Common Chimpanzee

Common chimpanzees are medium-sized apes. They are much smaller than gorillas. Male common chimps may grow to be 3 ft (91 cm) long and weigh 200 lb (90 kg). Females may grow to 2 ft 9 in (84 cm) in length and weigh 176 lb (80 kg). A chimp's length is the distance along its head and back when it is on all fours. Scientists measure chimps this way because these apes usually move on all fours rather than upright on their hind legs alone.

Common chimpanzees live in West and Central Africa. They can be found in **tropical** forests and on the **savannas**

– these are the grasslands of Africa where trees are few and widely scattered. Each day, the chimps spend about four hours traveling about looking for food. They feed on hundreds of different types of plants. Sometimes they even hunt and kill small mammals, such as monkeys and pigs. They also eat birds' eggs, honey, termites, and ants. A chimp gathers ants by poking a twig into an ant nest so that the ants crawl onto it. The chimp then pulls the twig out and sucks the ants off with its lips.

In almost half of the countries where the common chimpanzee was plentiful, it is now extinct or nearly extinct. Like the other apes, the common chimpanzee is in

Chimps that know each other often hug when they meet. Common chimps live in groups that can have as many as 150 members. These big groups contain smaller "parties" of three to six apes that stick together.

Common Chimpanzee

danger mainly because people are destroying its forest home. However, common chimps are also hunted for their meat, or captured alive to be sold. One of the reasons people buy chimps is to use them for scientific study. Chemicals are tested on them. Since chimps are so like us, what happens to a chimp is likely to be what would happen to a person.

Captive chimps have been taught to understand words, to use sign language, and to do complicated tasks. In the wild, away from humans, chimps have taught themselves how to make and use tools. This chimp is using a twig to gather ants.

Sadly, many young chimps are caught to be sold as pets. Others are captured to be sold into circuses, because people like watching the apes perform tricks.

Action is being taken to save the chimpanzee. Hunting chimpanzees is now illegal in many places, and people are trying hard to stop wild chimps from being bought and sold. Unfortunately, it is very difficult to prevent poachers from killing or catching chimpanzees, even in national parks. If the common chimpanzee is to survive outside zoos, more protected areas, such as national parks, must be set aside, and ways must be found to stop poachers.

Area where the pygmy chimpanzee can be found

In spite of its name, the pygmy chimpanzee is only a little smaller than the common chimpanzee.

Pygmy Chimpanzee

The pygmy chimpanzee, or bonobo, has a narrower chest, smaller teeth, longer legs, and shorter arms than the common chimpanzee. Its head is also smaller, with long side-whiskers, and its face is completely black instead of pink or brown-black. Males grow up to 2 ft 9 in (84 cm) long and weigh about 85 lb (39 kg). Females grow up to 2 ft 6 in (76 cm) long and weigh around 68 lb (31 kg).

The pygmy chimpanzee lives in the steamy, tropical **rainforests** of Zaire in Central Africa. Like the other great apes, it is in danger mainly because its forest home is being cut down. Also, local people hunt the chimp and use its meat for food, or they prepare medicines and magical charms from parts of its body. Sometimes young chimps are captured alive and sold as pets. So far, few pygmy chimps have been used for scientific study. It is now against the law to hunt or capture pygmy chimpanzees, but government officials find it hard to stop these practices.

Scientists do not know how many pygmy chimps are left, since there are still large areas of the ape's habitat that have not been searched. It is known, however, that Zaire's national parks contain few pygmy chimps. People are now working to set up a special national park in an area where pygmy chimps are known to live.

Grooming is an important part of life for pygmy chimps, as it is for common chimps. Apart from keeping fur clean, it helps the animals in a group to get to know one another and build close relationships.

Areas where the orangutan can be found

Unlike all other apes, the orangutan is a solitary animal, which means it lives alone.

Orangutan

The orangutan is a large ape with long, shaggy hair. Young orangutans have a bright orange coat, but when they become adults their coats usually turn maroon or dark

brown. Males weigh up to 200 lb (90 kg) and can grow to be nearly 5 ft (1.5 m) tall. Females are much smaller. They rarely weigh more than 100 lb (45 kg), and are usually shorter than 4 ft (1.2 m) tall. Both males and females have large, flabby pieces of skin on their necks. Males may also have flaps of skin on their cheeks. Sometimes males have moustaches and beards.

Years ago, orangutans lived in many parts of Asia, but today they can be found in the wild on only two islands in Southeast Asia: Borneo and Sumatra. These two groups of orangutans look a bit different from each other. Those from Sumatra are thinner and their coat has a lighter color. Bornean males have very large cheek flaps and throat flaps. They also have longer hair and a slightly longer face.

At night, orangutans make a leafy nest high up in the trees, about 30-70 ft (9-21 m) above the ground.

The orangutan spends more time in trees than any of the other great apes do.

Both groups make their home in tropical rainforests. There, trees can grow as tall as 150 ft (46 m). During the day, orangutans travel slowly from tree to tree in search of food. Over half of their diet is made up of fruit. They like to eat mangoes, figs, lychees (small round fruits that have a thin green skin), and rambutans (oval red fruits covered in soft spines). They also eat the young leaves and shoots of plants, tree bark, and sometimes eggs and small animals.

The orangutan faces the same threats as the other apes do. Its forest habitat is being cut down to harvest the valuable timber and to make room for farmland. In addition, hunters capture young orangutans to sell, even though it is against the law to catch or kill orangutans. The world's zoos have agreed not to buy any more orangutans, but some people still want them as pets.

Some steps have been taken to protect the orangutan. Large reserves have been set up in Borneo and Sumatra, where the animals can live more safely from people. Also, conservationists are making a special effort to help captive orangutans. Many of these animals have been rounded up and taken to centers where they are being taught skills that will allow them to live in the wild again.

The orangutan is in great danger. Only about 20,000 remain in the wild. Thousands of these great apes are believed to be dying each year. Unless their rainforest habitat is protected, orangutans will survive in only a few protected areas of Southeast Asia.

The word "orangutan" means "man of the forest." This name was given to the ape because sometimes it does indeed look like an old man.

This siamang gibbon has inflated its throat pouch. The pouch helps the gibbon to make a clear, loud call. All kinds of gibbons call to each other. These "songs" are some of the most beautiful sounds in the forests of Southeast Asia.

Gibbons

The gibbons are the smallest of the apes. Most kinds of gibbons measure just over 2 ft (61 cm) long and weigh up to 15 lb (7 kg) when fully grown. Females and males are similar in size and weight. There are nine kinds of gibbons. The largest gibbon is the siamang, while the agile and lar gibbons have the widest ranges. Each kind of gibbon has its own special markings, but all have very long arms.

The gibbons are found throughout Southeast Asia, including the islands of Indonesia. Over most of their range, they live in tropical rainforests, where the trees keep their leaves throughout the year. In some places, though, gibbons live in forests where the trees lose their leaves for part of the year.

Gibbons live in family groups. These usually consist of an adult male and female, and two to four of their offspring. Gibbons are mainly vegetarians, and feed on fruit and leaves. But some also eat insects.

Gibbons are generally less threatened by hunters than the great apes are. This is partly because gibbons are harder to find, and partly because they are not so popular as pets as chimps and orangutans are. Many kinds of gibbons are at risk, however. These include Kloss's gibbon, the Javan gibbon, and the black gibbon. The major threat facing gibbons is destruction of their habitat.

A lar gibbon makes it way through the trees. A gibbon's arms are as long as its legs and body put together.

Areas where Kloss's (brown) and the Javan (red) gibbons can be found

Kloss's gibbon is also known as the beeloh. It is black all over.

Kloss's & Javan Gibbons

Kloss's gibbon can be found only on the Mentawai Islands of Indonesia, where it lives in the tropical rainforest. The female is famous for her song. People have described it as the most beautiful sound they have heard a mammal make.

Kloss's gibbon is losing its forest habitat very quickly. Logging companies are felling huge numbers of trees for timber. They have even been given permission to cut trees in areas once set aside as reserves. There are also plans to move people to the Mentawais from other, more crowded parts of Indonesia. This will mean clearing more forest to make way for farmland. However, there are still areas of protected forest on the Mentawais. These need to be safeguarded if Kloss's gibbon is to survive.

The Javan gibbon also makes its home in Indonesia. It is silvery-gray in color and lives on the island of Java. Java is one of the most crowded places on Earth, and much of the gibbon's rainforest habitat has been destroyed to make way for homes and farms. Today only a tiny amount of the rainforest remains, and only part of it has been protected in areas such as the Gunung Halimun Reserve. There are probably less than 10,000 Javan gibbons left. If this ape is to survive, the reserves must continue to exist. If possible, they should be made larger.

A Javan gibbon leaps from branch to branch. Apes are related to monkeys. One big difference between the two is that apes do not have tails, and monkeys do.

Areas where the black gibbon can be found

A female black gibbon hangs from a branch. Gibbons usually travel by swinging through the trees using their arms.

Black Gibbon

The black gibbon lives in parts of Southeast Asia, including Vietnam and Laos. The male is black in color, sometimes with white cheeks; the female is a golden color, sometimes with black patches.

The black gibbon's forest habitat is being cut down to make room for people. The gibbon is also illegally hunted. There may be as few as 25,000 left. Scientists are not sure because there have been many wars in the area. This has made it difficult to study the gibbons.

Some governments are trying hard to protect wildlife. In Vietnam, for example, there are plans to set up national parks, where gibbons could live in peace. Vietnam and Laos are also trying to protect the rainforests by preventing people from selling the timber.

The biggest problem facing all the apes is the loss of their forest homes. Many forests are being destroyed for their valuable timber, and many are being cut down to make room for the growing number of people in the world. More safe areas need to be set aside so that these remarkable animals can continue to share our world.

The hoolock gibbon, another kind of gibbon that is endangered, peers out from its leafy perch.

Useful Addresses

For more information about apes and how you can help protect them, contact these organizations:

The Dian Fossey Gorilla Fund
45 Inverness Drive East
Englewood, CO 80112

International Primate Protection League
Box 766
Summerville, SC 29484

The Jane Goodall Institute
Box 41720
Tucson, AZ 85717

Orangutan Foundation International
822 South Wellesley Avenue
Los Angeles, CA 90049

U.S. Fish and Wildlife Service
Endangered Species and Habitat Conservation
400 Arlington Square
18th and C Streets NW
Washington, D.C. 20240

World Wildlife Fund
1250 24th Street NW
Washington, D.C. 20037

World Wildlife Fund Canada
90 Eglinton Avenue East
Suite 504
Toronto
Ontario M4P 2Z7

Further Reading

Amazing Monkeys Scott Steedman (New York: Knopf, 1991)

Endangered Wildlife of the World (New York: Marshall Cavendish Corporation, 1993)

Gorilla Rescue J. Bailey (Austin, TX: Raintree Steck-Vaughn, 1994)

Gorillas Paul H. Burgel and Manfred Hartwig (Minneapolis, MN: Carolrhoda, 1993)

Mountain Gorilla Michael Bright (New York: Gloucester, 1989)

Primates: Apes, Monkeys, Prosimians Thane Maynard (New York: Franklin Watts, 1994)

Saving Endangered Mammals: A Field Guide to Some of the Earth's Rarest Animals Thane Maynard (New York: Franklin Watts, 1992)

Wildlife of the World (New York: Marshall Cavendish Corporation, 1994)

Glossary

Adapt: To change to new conditions in order to survive.

Conservationist (Kon-ser-VAY-shun-ist): A person who protects and preserves the Earth's natural resources, such as animals, plants, and soil.

Extinct (Ex-TINKT): No longer living anywhere in the world.

Groom: To search an animal's coat and remove any insects, dirt, or knots of fur.

Habitat: The place where an animal lives. For example, the western lowland gorilla's habitat is the forest.

Mammal: A kind of animal that is warm-blooded and has a backbone. Most mammals are covered with fur or have hair. Females have glands that produce milk to feed their young.

Poaching: Illegal hunting.

Primate: A kind of mammal with human-like hands, eyes that face the front, and a well-developed brain. Monkeys, apes, and human beings are examples of primates.

Rainforest: A forest that has heavy rainfall all year.

Range: The area in the world in which a particular kind of animal can be found.

Reserve: Land that has been set aside for plants and animals to live in without being harmed.

Savanna: A wide open plain in a tropical area of the world. Grass is the main type of plant, and trees are few and widely scattered.

Silverback: An adult male gorilla. It gets its name from the silver-gray patch of hair on its back.

Troop: The name given to a group of gorillas.

Tropical: Having to do with or found in the tropics, the warm region of the Earth near the Equator. For example, a tropical rainforest.

Vegetarian: An animal that eats only plants and plant parts, such as seeds, nuts, and fruit.

Index

agile gibbon 24

black gibbon 25, 28-29
Bwindi Forest 10, 11

chimpanzees 4, 14-19, 25
common chimpanzee 14-17, 18

eastern lowland gorilla 5, 12-13
Eastern Zaire Gorilla Conservation
   Project 13

gibbons 4, 24-29
gorillas 4, 5-13, 14
Gunung Halimun Reserve 27

hoolock gibbon 29

Javan gibbon 25, 26-27

Karisoke Research Centre 11
Kigezi Gorilla Sanctuary 11
Kloss's gibbon 25, 26-27

lar gibbon 24, 25

mountain gorilla 5, 10-11, 12, 13

orangutan 4, 20-23, 25

pygmy chimpanzee 18-19

siamang 24
silverback 6, 7

Virunga Mountains 10, 11
Virunga National Park 11

western lowland gorilla 5, 6, 8-9